Steam Locomotives of Indian Railways

About the Book

Published By:- Er. Twahir Alam.

Copyright © 2019 Er. Twahir Alam

All rights reserved.

ISBN: 9781709363405

Edition:-

First Edition:- November, 2019.

PUBLISHERS' Note

This publication is being sold on the condition that the information, comments and views that are contained in it are merely for guidance and reference.

No part of this publication may be reproduced, in any form or by any means – electronic, mechanical, recording, photocopying or otherwise without the prior permission from the publishers.

PUBLISHERS'

DEDICATION

The book is dedicated to all men and women of the Indian railways who by their sheer hard work and dedication has not only kept the nation's pride running but taken it to new heights.

CONTENTS

1. Introduction — 1
2. Steam Traction — 3
3. Nomenclature — 5
4. Types of Locomotives — 6
5. Broad Gauge Locomotives — 10
6. Meter Gauge Locomotives — 56
7. Narrow Gauge Locomotives — 61
8. Production Units — 67
9. Steam Locomotive Sheds — 86
10. Conclusion — 102

ABOUT THE AUTHOR

Er. Twahir Alam is a civil servant, working for the Government of Assam in Assam Civil Service. He is an Electrical Engineer having passed out from Assam Engineering College and has lots of experience working in Engineering field. He has worked as an Executive for Crompton Greaves in Delhi and also for Assam Power Distribution Company Limited as an Assistant Manager. He also worked as an Excise Inspector in the Excise Department. He is a train enthusiast also and has a section of his YouTube channel dedicated to Indian Railways.

INTRODUCTION

The Indian Railways is righty regarded as the lifeline of the nation as it is the true mover of India. The story of Railways began in India on the 16th of April 1853, when the very first train of the Indian Railways was hauled from Bombay's Bori Bunder railway station to Thane. The train by hauled by a combination of three steam locomotives named Sahib, Sultan and Sindh, over a distance of about 34 kms. The train carried 400 passengers in 14 carriages. The Railway has grown leaps and bounds since then and has spread its network throughout the country. Today has grown to become the fourth largest railway network by size in the World, with total track of 67,368 kms. It now operates about 13,000 passenger trains in its network of about 7,349 stations and 9,200 goods trains daily in its network.

The evolution of the Indian Railways is one of the most interesting episode of modern Indian history but sadly one which has received very little attention. Yet for technologists, engineers and railway enthusiasts like me, it is a breath taking journey and one which has to be studied in detail and told to the World.

The journey and growth of the Indian Railways has not been a smooth or linear one, but one filled with success, failures and lots of experimentations. It is this which makes the study of its history even more fascinating. Thus, we can find many technologies and policies being used and tried to make the Railways more efficient and their success has contributed greatly to the growth of Indian Railways. Steam traction was used in Indian Railways since its inception and it was the technology which was used by the Railways as it grew and spread itself to the entire nation. The very sight of them created a sensation as among the people, as they watched a revolution taking place in the way they travelled. The steam locomotives were thus the pioneers of rail travel in India, as most trains were operated by them. While DC Electric trains were introduced by the 1920s, but they were very limited in their area of operation. Thus, steam locomotives became synonymous to rail travel and a kind of romanticism grew with them but with time its inherent limitations was apparent and soon they were replaced by diesel and electric locomotives. Today steam locomotives do not operate on the main lines of the Indian Railways and are used to haul heritage trains and on heritage railways only but the romanticism which they generated would remain forever.

In this book, we would trace the journey of the steam locomotives and look into the types and characteristics of the locomotives used in India.

STEAM TRACTION

The steam locomotives were the first type of locomotives to be used to haul trains on the Indian Railways. The steam locomotives were thus the pioneer locomotives in India and they were the ones who introduced the Railways to India. Most of the early steam locomotives used in India, were imported from UK but they served the Railways well. As the Railway network in India grew from a small area to cover the whole country, it the steam locomotives which helped who propelled this expansion.

In a steam engine, a boiler produces steam which in turn moves reciprocating piston connected to the locomotive's main wheels and thus moves it forward. Wood, coal or oil is burned in the boiler to produce steam. As the locomotives require huge quantities of fuel and water, they have to be carried with it and this may be done inside the locomotive itself or in a wagon attached to it.

The steam locomotive consists of the boiler, the steam circuit, the running gear and the chassis. As these were used at time when everything was mechanical, these locomotives tended to demand a lot from the users and were vastly different from today's locomotives. The physical demand on the drivers and operators was as much as was the mental demand. Apart from the tiring work of shuffling the coal, the operation of various levers and other equipments needed lot of physical strength.

While the highly demanding steam locomotives, were very tiring on the workmen but it to them developing very close and emotional bonds with the locomotive. This led to growth of the romantic tales associated with them and one which is so strong that even after decades of being obsolete, they are still used to run trains. An honor which was not given to any other retired locomotive types like the DC locomotives.

History of Steam

The first commercially successful steam locomotive was built by John Blenkinsop in 1812 and Locomotion 1, built by George Stephenson became the first steam locomotive to haul passengers on a public railway i.e. on the Stockton and Darlington Railway in 1825.

The first steam locomotive ran on the Red Hill Railway in Madras in 1837 and was used to haul granite for road building. The line ran from Red Hills to the Chintadripet Bridge and was not part of any Railways. The steam locomotive became the pride of the Indian Railways with time. While DC traction was introduced in 1920s, it was very much restricted to a few zones and thus it was on these steam locomotives that the Indian Railways was able to go to all over the country.

NOMENCLATURE

Indian Railways did not have any standard nomenclature system for the steam locomotives and different nomenclature was used for different designs of locomotives.

TYPES OF LOCOMOTIVES USED

The Indian Railways used over 30,000 steam locomotives of various classes and types in the 150 years of their operation. The various types of Steam locomotives used in Indian Railways are as follows:-

I. Broad Gauge Locomotives:-
1. Custom Build Type:-
2. BESA Standard Locomotives:-
   a. Class SP.
   b. Class SG.
   c. Class PT.
   d. Class HP.
   e. Class AP.
   f. Class HG.
   g. Class HT.
3. IRS Locomotives:-
   a. Class XA.
   b. Class XB.
   c. Class XC.
   d. Class XD.
   e. Class XE.
   f. Class XF.
   g. Class XG.
   h. Class XH.
   i. Class XP.

       j. Class XS.

       k. Class XT.

4. Post World War II Locomotives:-

       a. Class WP.

       b. Class WG.

       c. Class WL.

       d. Class WM.

       e. Class WT.

       f. Class WU.

       g. Class WV.

II.   Meter Gauge Locomotives:-

1. Nilgiri Mountain Railway X class

2. BESA designs:

    a)   Passenger (4-6-0)

    b)   Mixed (4-6-0)

    c)   Goods (4-8-0)

    d)   Tank (2-6-2T)

3. Wartime designs:

    a)   Class MAWD.

    b)   Class MWGX.

4. Indian Railway Standards designs

    a)   Class YA.

    b)   Class YB.

    c)   Class YC.

- d) Class YD.
- e) Class YE.
- f) Class YF.
- g) Class YK.
- h) Class YL.
- i) Class YT.
- j) Class YG.
- k) Class YP.

III. Narrow Gauge (2 Ft 6 In)
  1. Barsi Light Railway:
     - a) Class A.
     - b) Class B.
     - c) Class C.
     - d) Class D.
     - e) Class E.
     - f) Class F.
     - g) Class G.
  2. Indian Railway Standards:
     - a) Class ZA.
     - b) Class ZB.
     - c) Class ZC.
     - d) Class ZD.
     - e) Class ZE.
     - f) Class ZF.

IV. Narrow Gauge (2 Ft)
1. Darjeeling Himalayan Railway:-
   a) DHR A Class.
   b) DHR B Class.
   c) DHR C Class.
   d) DHR D Class.
2. Indian Railway Standards:-
   a) QA.
   b) QB.
   c) QC.

Broad Gauge Locomotives

Custom Build Locomotives

The first steam engines which were used in India, were all custom built ones, as no standards for locomotive was established at that time. Thus, each customer had a different type of locomotive which were built as per their requirements and specifications. So, during the early days of the steam locomotives, we see a large of locomotives being used in India and each type was a unique one. These were mainly manufactured in UK and imported to India. While the custom built locomotives ensured that each Railway had a locomotive which was best suited for its use, purpose and terrain but it had them very costly and also the production time was very long.

In the 1890s, as British manufacturers were busy, a number of locomotive was also imported from the US and Germany. The locomotives of this type used by the different railways of India are as follows:-

Bengal Nagpur Railway

The Bengal Nagpur Railway also imported many steam locomotives and they were:-

1. Class F.
2. Class GM.
3. Class HSG.
4. Class M.
5. Class N.
6. Class NM.
7. Class P.

The Class F, was based on the 0-6-0 configuration of locomotive and the Class GM was based on the 2-6-0 configuration. The Class M were based on 4-6-2 configuration. Most locomotives used by the Railway were built by the Beyer, Peacock and Company, UK.

Class HSG

These BNR Class HSG was actually the 2-8-0+0-8-2 Garratt locomotive and was the first successful class of Garratts. A total of only two of these locomotives were imported from Beyer, Peacock and Company. They were used to haul heavy trains in the ghat section and were quite successful in doing so. They were mostly run together at the Chakradharpur-Jharsuguda section

They were used till 1969, when the lines they operated were electrified. They were then stationed at Kharagpur workshops but now both of them have been scrapped.

Factsheet:-
- Fuel Type:- Coal.
- Manufacturer:- Beyer, Peacock and Company, UK.
- Production Period:- 1925.
- Numbers Produced:- 2.
- Whyte:- 2-8-0+0-8-2.
- Rated Power Output:- HP
- Top (Rated) Speed:-   Kmph.
- Weight:- 180.5 Tons.
- Serial Numbers:- 38681 and 38692.

Class N

These Class N like the Class HSG locomotives were also a Garratt locomotive and was of the 4-8-0+0-8-4 configuration. A total of 16 of these locomotives were imported from Beyer, Peacock and Company, UK in 1929. The success of the HSG Class, led to the development of this class of locomotive and at the time of their import, they were the largest locomotive in India and had the largest water capacity of any Garratt.

The configuration of these locomotives were effectively two 4-8-0 locomotives operating back to back. They were also called Double Mastodon.

They were mostly used in the Chakradharpur-Jharsuguda and Anara-Tatanagar sections and when the sections were electrified, they were in sent to Rourkela. These locomotives were very heavy and so required heavy rails. They continued to serve the Indian Railways for decades and were finally retired in 1970. Today, two of them numbered 811 and 815, have been preserved at National Rail Museum, Delhi and at Kharagpur Workshop respectively.

Factsheet:-

- Fuel Type:- Coal.
- Manufacturer:- Beyer, Peacock and Company, UK.
- Production Period:- 1929.
- Numbers Produced:- 16.
- Whyte:- 4-8-0+0-8-4

- Tractive Effort:- 309.84 kN.
- Top (Rated) Speed:- 72 Kmph.
- Weight:- 234 Tons.
- Serial Numbers:- 38810 to 38825.

Class NM

The NM Class of steam locomotives were modified types of the N Class of Garratt locomotives. A total of 10 such locomotives were imported from Beyer, Peacock and Company in 1931. They were used on the Bilaspur-Katni line and Anuppur-Chirmiri branch line for hauling coal. They did not remain in service for long and were retired very early in 1960.

Factsheet:-
- Fuel Type:- Coal.
- Manufacturer:- Beyer, Peacock and Company, UK.
- Production Period:- 1931.
- Numbers Produced:- 10.
- Whyte:- 4-8-0+0-8-4.
- Tractive Effort:- 309.86 kN.
- Top (Rated) Speed:-   Kmph.
- Weight:- 204.15 Tons.
- Serial Numbers:- 38826 to 38835.

Class P

The Class P type of locomotive of the Bengal Nagpur Railway was also another locomotive of Garratt type. The locomotive was developed from the Class N and Class NM type of locomotives and were specially designed for the Anuppur-Chirmiri section. This section had a very severe curvature and so required a locomotive with a trailing bogie. The 4-8-2 configured locomotives were usually called as Mountain but these were call Double Mountain as they consisted of a pair of engines. The locomotive is in fact a combination of two engines combined to give more power and higher traction.

Factsheet:-

- Fuel Type:- Coal.
- Manufacturer:- Beyer, Peacock and Company, UK.
- Production Period:- 1939.
- Numbers Produced:- 4.
- Whyte:- 4-8-2+2-8-4.
- Tractive Effort:-   kN.
- Top (Rated) Speed:- 309.84 Kmph.
- Weight:- 230 Tons.
- Serial Numbers:- 38855 to 38858.

Bombay, Baroda and Central India Railway

The Bombay, Baroda and Central Indian Railway, was the predecessor of the current Western Railways zone of the Indian Railways. It too imported a number of custom built steam locomotives.

The locomotives used it were the Class P, which was a locomotive with a 4-6-2 wheel configuration. The Class A, was a locomotive with a 2-4-0T configuration and was based at the Parel shed. The Class U36, was a 0-4-2 configuration locomotive and was used to haul suburban trains in Bombay. The Class D1 was a 4-4-0 configuration locomotive was named as "Princess May". The Class M was a 4-6-2 configured locomotive.

Eastern Indian Railway Company

The Eastern Indian Railway Company, was the pioneer Railway which introduced the Eastern India to the Railways and it too operated a number of custom built steam locomotives which included the Class CT, was a 0-6-4T configured locomotive and it was later converted into a super heater. The EIR Class was a 4-6-0 and the EIG Class, was a 2-2-2T configuration locomotive. They are the World's Oldest working locomotives. The first two were named as the Express and Fairy Queen and was rebuilt by Perambur Loco Works.

The Fairy Queen EIR 22, is today the oldest running locomotive of the World. While it was built in 1854 in UK, its first commercial run was on the 15th of August 1855 from Howrah. The locomotive was retired in 1908 but was revived and made operational in 1997. The locomotive was looted in 2011, in Delhi and remained out of service for many years. It has been made operational again after a gap of six years.

Another locomotive of the same class bearing the serial number EIR 21, also known as Express was revived in 2010. It was built in 1855 by Kitson, Thompson & Hewitson of England and remained in operation till 1909. It was then kept as an exhibit at Jamalpur Workshop and Howrah station for over a hundred years. The hundred years of exposure to all the elements had taken a heavy toll on the locomotive and no hope left for it, but only due to the sheer will and determination of the staff of the Loco Workshop, Perambur that the locomotive was revived. Today, it is used to haul special trains.

Great Indian Peninsular Railway (GIPR)

The GIPR too operated a number of custom built steam locomotives and the classes Y1, Y2, Y3, and Y4 were 0-8-4T configured locomotives and were used as bankers on the Thul Ghat. The Class F and F3, were 2-6-0 configured locomotives while the Class J1, was a 0-6-0 configured locomotive. The Class D4 and D5, were both 4-6-0 configured locomotives and the D5 was a passenger locomotive.

The first commercial train in India which ran between Bombay and Thane on the 16th of April 1853, was hauled by three locomotives named as Sahib, Sultan and Sindh. These were 2-4-0 configured tender locomotives which were produced by Vulcan Foundry. They were part of a batch of eight such locomotives which was imported into India. Sadly today none of these three locomotive have survived.

The Class E1, was a 4-4-2 configured locomotive and was built by the North British Locomotive Company in 1907-08 and later rebuilt as a super heater in 1925. The Class T, was a tank locomotive and was used to haul suburban train in Bombay. The Class Y, was a 2-8-4T configured locomotives.

The Crane Tank engine, was a very unique locomotive of the Railways and has been preserved at the National Rail Museum at New Delhi.

Other Railways

The Madras and Southern Mahratta Railway used a number of custom built locomotives which included the M&SM Class V, Class BTC and Class T. The M&SM Class V, was a 4-4-0 configured locomotive and has been preserved. The Class BTA, is a 2-6-4T configured locomotive and was based on BESA specifications and the Class T, was a 0-4-2 configuration locomotive and is now preserved in Madras.

The Thomason, was one the first steam locomotives of India and it arrived in India in about 1951. It was used during the construction of the Sonali Aquaduct near Delhi. It was a six wheeled well tank engine, with a configuration of 2-2-2. It was manufactured by E.B. Wilson.

The FALKLAND, was another of the pioneer steam locomotives of Indian Railways. It was brought in 1852 and was a construction locomotive. It was used to construct the line between Boribunder and Thane. It had a configuration of 0-4-0 and was later transferred to the GIPR. It too was manufactured by E.B. Wilson.

The Nizam's Guaranteed State Railway, used the Class A locomotive and had a 2-6-0T configuration. One of them is now preserved at the National Rail Museum at New Delhi.

The Oudh and Rohilkhand Railway, used the Class B26, which was a 0-6-0 configured locomotive and one of them is also preserved at the National Rail Museum at New Delhi.

The North Western Frontier Railway, used a number of such custom built locomotives and they were the EM Class, GAS Class, P Class, E1 Class and the N1 Class. While the Class P was a 2-4-0, the Class E1 was a 4-4-2 and the Class N1 was a 4-8-0 configured locomotive. The GAS is a 2-6-2+2-6-2 configured locomotive and was built by Garratt in 1925 and was retired by 1937. The EM Class was a 4-4-2 configured locomotive and is now preserved at the National Rail Museum.

A large number of other such customized locomotives were used all over India and they were as follows. They were the Class NA2, the Class B, which was a 2-6-0 and the Class E which was a 2-4-0 configured locomotive. The Class G, was a 2-6-0 configured goods locomotive and the Class Y2 was a 2-8-2T configured locomotive. The Class PTC, was 2-6-4T passenger locomotive and was owned by the Northern Railway. The Class F, was a 2-8-2 configured locomotive of the Central Railways and was built by Nasmyth Wilson from 1926 to 1950.

The Phoenix, is a 0-4-0T configured locomotive, while the Ramgotty is a 2-2-0T configured locomotive, which was later converted to a broad gauge locomotive. Both these locomotives are now preserved at the National Rail Museum.

BESA Standard Locomotives

The increase in the network of Railways with time and requirements, led to increased demand for locomotives. While the custom built locomotives had many advantages such as that each Railways has locomotives built as per their need, terrain and conditions but required lot of time to manufacture. The increased demand of locomotives warranted that the manufacturing time be reduced and so in 1903, standard designs for locomotives were made by the British Engineering Standards Committee for the Indian Railways. A 4-4-0 locomotive was developed for passenger operations while a 0-6-0 was developed for goods operations. The designs were adopted fully by the state railways but were modified by the companies for their uses.

The locomotives which were designed and produced in this series are as follows:-

a. Class SP: Standard passenger (4-4-0).
b. Class SG: Standard goods (0-6-0).
c. Class PT: Passenger tank (2-6-4T).
d. Class HP: Heavy passenger (4-6-0).
e. Class AP: Atlantic passenger (4-4-2).
f. Class HG: Heavy goods (2-8-0)
g. Class HT: Heavy tank (2-8-2T).

When superheating was introduced, the letter S was suffixed to the Class to devote it e.g. HPS and if the locomotive was a converted from saturated to superheated one then the letter C was suffixed e.g. SPC, SGC.

In 1949-50, the last order for locomotives based on these designs were given to Vulcan Foundry. They were given as the new WP locomotive was not thoroughly tested as that time and the failure of the IRS Standar made the Railways go with a proven and successful design. The order was for 84 modified locomotives and they were later designated as the HPS 1 and HPS 2 Class.

Class SG

The SG Class of steam locomotives were the first locomotives of the new BESA design and were goods locomotives. They were manufactured in UK by Vulcan Foundry, North British Locomotive Company and Robert Stephenson and Hawthorns from 1905 to 1930s and were the first standard goods locomotive of India. They were quite efficient and a total of 552 of them were produced and imported. They were initially used to haul passenger mail trains and were used till the end of the steam age in India.

A total of three variants of this Class was produced, while the first one was a goods locomotive, the second one named SGC2 was used to haul passenger and mail trains also. The third one named SGC3, was retrofitted with super heaters and was the only locomotive in the World to use the Lentz rotary valve gear with inside cylinders.

They were used by the, Eastern Bengal Railway, EIR and Oudh and Rohilkhand Railway and later by Easter and Northern Railway zone of Indian Railways. All of them have been retired and scrapped.

Factsheet:-
- Fuel Type:- Coal.
- Manufacturer:- Vulcan Foundry, North British Locomotive Company and Robert Stephenson and Hawthorns.
- Production Period:- 1905-1930s.
- Numbers Produced:- 552.

- Whyte:- 0-6-0.
- Tractive Effort:- kN.
- Top (Rated) Speed:- 29 Kmph.

Indian Railway Standards Locomotives

The increase in coal prices after the end of the First World War, made the running of locomotives with imported coal to be less economical but the BESA standard locomotives were not designed to work with the low quality Indian coal. At the same time, a number of railways carried a number of experiments to build locomotives which were capable of burning Indian coal. This was done by using a wider firebox and 4-6-2 configuration for passenger and 2-8-2 configuration for goods locomotives. These experiments were sadly very limited and had no impact on development of Indian Locomotives.

In 1919, to achieve the same goal of building locomotives to run on Indian grade coal and also to find replacement for the old BESA standards, the Locomotives Standards Committee was formed. The Committee brought out new standards which became the basis of the new class of locomotives which were designated as the Indian Railway Standards locomotives. The Committee recommended a total of five designs for broad gauge railways based on axle loads and duties. A number of other designs were also derived from these standards.

The locomotives which were designed and produced in this series are as follows:-

    a. Class XA.
    b. Class XB.
    c. Class XC.
    d. Class XD.

e. Class XE.
f. Class XF.
g. Class XG.
h. Class XH.
i. Class XP.
j. Class XS.
k. Class XT.

The five locomotives which were originally designed from the standards were found to have very poor and sluggish performance. While the XA, XD and XE did achieve some success but as a whole they were a failure. So they were soon taken off mainline duties and were replaced by the older BESA design locomotives.

Class XA

The Class XA, was a 4-6-2 configured passenger locomotive and was a light axle load locomotive. The locomotive was built by Vulcan Foundry in Newton-le-Willows, Lancashire, England from 1929 to 1935. A total of 113 of these locomotives were imported to India and after partition, 37 of them were given to Pakistan and the rest 76 remained in India.

The XA Class was the only locomotive the original five designs to achieve some success. All of them have been retired and most of them have been scrapped except two with serial numbers 22022 and 22046, which have been preserved.

Factsheet:-

- Fuel Type:- Coal.
- Manufacturer:- Vulcan Foundry, UK.
- Production Period:- 1929-35.
- Numbers Produced:- 113.
- Whyte:- 4-6-2.
- Tractive Effort:- 93.23 kN.
- Top (Rated) Speed:- 29 Kmph.
- Weight:- 42.1 Tons.
- Serial Numbers:- 22001 to 22076.

Class XB

The Class XB, was a 4-6-2 configured light passenger locomotive. The locomotive was designed by M/s Rendel Palmer and Tritton, with help from British Engineering Standards Association (BESA). It was the most advanced locomotive of its time in Britain and was produced by Vulcan Foundry Armstrong Whitworth and North British Locomotive Company from 1927 to 1936. The locomotives had American 3-point suspension with compensating levers for indifferent tracks in India. They proved to be quite successful in India and were used till the 1980s when steam locomotives were phased out but they had a number of initial problems. It used to haul trains like the West Coast Express and Darjeeling Mail.

The locomotives were found to be prone to fractures. The coupling rod and the tubeplates failed frequently and required large number of changes. To overcome the problems, a leading and trailing bogie with stiffer side springs and better damping was used and it was successful to remove many of its problems. XB class were found to have very poor performance.

The XB Class was involved in a major accident in 1937 in Bihar. A total of more than a hundred person were killed in the incident and the locomotive's speed was restricted to 45 kmph from then. The Committee of Experts, who enquired into the accident, found three design flaws with the locomotive and this was found the affect the front bogie, hind truck and the drawgrear.

The number of derailments, involving them and the XC Class, led to the reduction of their operational speed limit from 60 kmph to 45 kmph and finally to their withdrawal from mail line operations. Today none of the XB Class locomotive has been preserved in India, as all have been retired and scrapped in 1983. Only one is preserved by Pakistan Railways in Lahore.

Factsheet:-
- Fuel Type:- Coal.
- Manufacturer:- Vulcan Foundry Armstrong Whitworth and North British Locomotive Company.
- Production Period:- 1927 to 1936.
- Numbers Produced:- 99.
- Whyte:- 4-6-2.
- Tractive Effort:- 119.03 kN.
- Top (Rated) Speed:-.
- Weight:- 55.3 Tons.

Class XC

The Class XC, was also a 4-6-2 configured passenger locomotive but was suited for heavier axle loads. A total of 72 of this class of locomotive was produced in between 1928 to 1931 by William Beardmore & Co and Vulcan Foundry.

The XC Class along with the XB Class locomotive were not only found to be very poor in performance but they also were involved in a number of accidents. This led to their speed reduction from 60 kmph to 45 kmph and to their eventual withdrawal from mainline duties.

The locomotives were divided between Indian and Pakistan Railways during partition and India got 50, while the rest when to Pakistan. Sadly no locomotive of this class is preserved today.

Factsheet:-
- Fuel Type:- Coal.
- Manufacturer:- William Beardmore & Co and Vulcan Foundry.
- Production Period:- 1928 to 1931.
- Numbers Produced:- 72.
- Whyte:- 4-6-2.
- Tractive Effort:- .
- Top (Rated) Speed:-.

Class XD

The Class XD, was a 2-8-2 configured goods locomotive and was suited for light goods operations. It had an axle load of 17 tons. The locomotives were initially ordered by the Madras and Southern Maratta Railway. The locomotive is equipped with vacuum brakes which are also operated by steam. These locomotives achieved moderate success with their performance.

Factsheet:-

- Fuel Type:- Coal.
- Manufacturer:- Vulcan Foundry, UK.
- Production Period:- 1940.
- Numbers Produced:- 35.
- Whyte:- 2-8-2.
- Axle Load:- 17 Tons.

Class XE

The Class XE, was a 2-8-2 configured goods locomotive and was suited for heavy goods operations. It had an axle load of 22.5 tons. Today five of them are preserved all over India including one by MP Electricity Board and UP Cement Corporation. A total of 51 of them were built at Vulcan Foundry and were the first export order from it after the Second World War. The boiler had four arch tubes and wide round-topped firebox with combustion chamber. It had a water capacity of 6,000 gallons and coal capacity of 14 tons. These locomotives achieved moderate success with their performance.

Factsheet:-

- Type:- Coal.
- Manufacturer:- William Beardmore & Co and Vulcan Foundry.
- Production Period:- 1928 to 1945.
- Numbers Produced:- 93.
- Whyte:- 2-8-2.
- Tractive Effort:- 226.5 kN.
- Weight:- 87 Tons.
- Axle Load:- 22.5 Tons.
- Serial Numbers:- 22501 to 22593.

Class XF

The Class XF, was a 0-8-0 light shunting locomotive with an axle load of 18 Tons.

Factsheet:-
- Fuel Type:- Coal.
- Manufacturer:- .
- Production Period:- 1928.
- Numbers Produced:- 6.
- Whyte:- 0-8-0.
- Axle Load:- 18 Tons.

Class XG

The Class XF, was a 0-8-0 heavy shunting locomotive with an axle load of 23 Tons.

Factsheet:-

- Fuel Type:- Coal.
- Manufacturer:- .
- Production Period:- 1928.
- Numbers Produced:- 3.
- Whyte:- 0-8-0.
- Axle Load:- 23 Tons.

Class XH

The Class XH, was 4 cylinder 2-8-2 experimental shunting locomotive design, with an axle load of 28 Tons. No locomotive of this class was ever built.

Class XP

The Class XP, was an experimental locomotive class with 4-6-2 configuration and an axle load of 18.5 Tons. Only two of this class was produced by Vulcan Foundry for the Great Indian Peninsular Railway in 1935. They were designed for heavy performance parameters and these were very ambitious at the time of their design. This Class was intended to have the power of the XC Class Heavy Passenger Locomotive with the operating area of the light load XB Class Locomotives. The basis of the locomotive was the XB Class Locomotive but had higher boiler pressure and tractive effort.

The XP Class had an axle load of 18.7 tons and working pressure of 210 lbs per sq. in. which was higher than the 180 lbs of the IRS class. This was achieved by using improved boilers. Its tractive effort was also much higher than the XB and XC Class locomotives. They were fitted with Caprotti valve gears and roller bearings.

The two locomotives entered service with the Great Indian Peninsular Railway in 1937 with the serial numbers 3100 and 3101. They were named as the King George and the Queen Elizabeth. In November 1951, the GIPR incorporated into the Central Railway and the locomotives were renumbered as 22599 and 22600. They were withdrawn in 1970 and now have been scrapped.

The Locomotive Class later became basis of the WP Class Locomotive.

Factsheet:-
- Fuel Type:- Coal.
- Manufacturer:- Vulcan Foundry.
- Production Period:- 1935.
- Numbers Produced:- 2.
- Whyte:- 4-6-2.
- Tractive Effort:- 147.04 kN.
- Top (Rated) Speed:-.
- Weight:- 99 Tons.
- Axle Load:- 18.7 tons.
- Serial Numbers:- 22599 & 22600.

Class XS

The Class XP, was also an experimental locomotive class with 4-6-2 configuration but had a higher axle load of 21.5 Tons. The Locomotive was manufactured by Vulcan Foundry in 1930 and a total of 4 of such locomotives were built. The Class is divided into two sub classes i.e. XS 1 and XS 2 and two locomotive of each sub class was built. The locomotives served with the North Western Railway and all were transferred to Pakistan Railway after partition. While the dimensions of the two sub classes were similar, a number of differences existed in the internal mechanisms. All four locomotives were deployed between Peshawar and Lahore and Lahore and Karachi.

Factsheet:-
- Fuel Type:- Coal.
- Manufacturer:- Vulcan Foundry.
- Production Period:- 1930.
- Numbers Produced:- 4.
- Whyte:- 4-6-2.
- Tractive Effort:- 153.02 kN.
- Top (Rated) Speed:-.
- Weight:- 121 Tons.
- Axle Load:- .
- Serial Numbers:- 760 & 761.

Class XT

The Class XT was a light tank locomotive with 0-4-2T configuration and an axle load of 15 tons only. The Ajmer workshop of BB&CI Railways also produced a number of them. A total of twenty of them were produced by the workshop and the first batch of 10 was produced in February 1947 and was handed over to the North Western Railways. The remainder 10 was produced in 1947-50 and was handed to Eastern Punjab Railway. The Eastern Punjab Railway, was the Indian part of the North Western Railways after partition.

Factsheet:-

- Fuel Type:- Coal.
- Manufacturer:- .
- Production Period:- 1929.
- Numbers Produced:- 77.
- Whyte:- 0-4-2T.

World War II and Post World War II Locomotives

The period of World War II saw a number of locomotives being imported into India from the United States and Canada. They were classified as the AWD and CWWD Class Locomotives. The Class AWC was also used at that time and was built by Baldwin Locomotive Works by adapting the USATC S160 Class locomotive design for India. A total of 60 locomotive of this Class was built.

While a number of locomotives were designed during this period but all of them entered service only after the War. The Classes of Locomotives built after the Second World War are as follows:-

1. Class WP.
2. Class WG.
3. Class WL (1st).
4. Class WL (2nd).
5. Class WM.
6. Class WT.
7. Class WU.
8. Class WV.
9. Class WW.

The first 16 of this Class was imported from Baldwin Locomotive Works, USA in 1947 and were classed as WP/P. They were numbered from 7200 to 7215 and were prototype locomotives. While at that time the order of locomotive was 100, only 16 of the new locomotive was ordered, as the Railways wanted to test them thoroughly before placing bulk orders. The Railways instead went with the tried and successful design of the HPS Locomotive. 84 modified HPS Class locomotives were ordered from Vulcan Foundry in 1949. These locomotives became the HPS 1 and HPS 2 Class.

The locomotives were a huge success and soon earned great reputation for their performance and riding comfort. This led to the Railways placing bulk order for them with Baldwin, Canadian Locomotive Company (CLC) and Montreal Locomotive Works. The first main production batch of 300 locomotive was manufactured by them in 1949. They were numbered from 7216 to 7515. A total of 100 locomotive was built by Canadian Locomotive Company in 1955-56. Another 60 was produced by Fabryka Lokomotyw, of Chrzanów, Poland, and Lokomotivfabrik Floridsdorf of Vienna, Austria in 1957-59. The next batch of 259 was produced in India by CLW from 1963 and 1966. They were 5 tons heavier and were classified as the WP/Is.

The WP Class locomotives, were the crack locomotives of the Indian Railways in the 1960s and 1970s and hauled some of Indian Railway's most prestigious trains like the Taj Express, Grand Trunk Express, Howrah-Madras Mail, Frontier Mail and the Air Conditioned Expresses. The

Class WP

The WP Class of locomotive was a 4-6-2 configured passenge[r] locomotive. The locomotive marked the change of classification code 'X' to 'W'. A total of about 755 of them were manufactured by them 1947 to 1967.

The WP Class was the outcome of years of research and deve[lopment] into building more efficient and power locomotives and were built wi[th] goal to produce them in India at later stage. They were specially desi[gned] low calorie and high ash Indian coal by Railway designers in India. [At] the time of their design, most locomotives had an axle load of about 2[2], they were designed with a lower axle load of 18.5 tons. This was don[e to] improve their reliability. They were designed to have the horse powe[r] to the XC Class of locomotives of IRS standards but without the performance issues.

The firebox was increased to enable them to use low quality c[oal,] bar frames were also introduced. While the weight of the locomotive[s was] more than the XC Class, the even distribution of weight led to the ma[ximum] axle load of 18.5 tons was achieved. While its tender was designed t[o carry] 15 tons of coal and 6,000 gallons of water, it was reduced to 5,550 g[allons to] keep the axle load within limit. The locomotive also had a semi-strea[mlined] bullet-tip shaped nose in front of the smokebox but it was later foun[d to not] be of much use. It was never removed from the locomotive.

Class WP

The WP Class of locomotive was a 4-6-2 configured passenger locomotive. The locomotive marked the change of classification code from 'X' to 'W'. A total of about 755 of them were manufactured by them from 1947 to 1967.

The WP Class was the outcome of years of research and development into building more efficient and power locomotives and were built with the goal to produce them in India at later stage. They were specially designed for low calorie and high ash Indian coal by Railway designers in India. While at the time of their design, most locomotives had an axle load of about 20 tons, they were designed with a lower axle load of 18.5 tons. This was done to so improve their reliability. They were designed to have the horse power equal to the XC Class of locomotives of IRS standards but without the performance issues.

The firebox was increased to enable them to use low quality coal and bar frames were also introduced. While the weight of the locomotives was more than the XC Class, the even distribution of weight led to the maximum axle load of 18.5 tons was achieved. While its tender was designed to carry 15 tons of coal and 6,000 gallons of water, it was reduced to 5,550 gallons to keep the axle load within limit. The locomotive also had a semi-streamlined bullet-tip shaped nose in front of the smokebox but it was later found not to be of much use. It was never removed from the locomotive.

The first 16 of this Class was imported from Baldwin Locomotive Works, USA in 1947 and were classed as WP/P. They were numbered from 7200 to 7215 and were prototype locomotives. While at that time the order of locomotive was 100, only 16 of the new locomotive was ordered, as the Railways wanted to test them thoroughly before placing bulk orders. The Railways instead went with the tried and successful design of the HPS Locomotive. 84 modified HPS Class locomotives were ordered from Vulcan Foundry in 1949. These locomotives became the HPS 1 and HPS 2 Class.

The locomotives were a huge success and soon earned great reputation for their performance and riding comfort. This led to the Railways placing bulk order for them with Baldwin, Canadian Locomotive Company (CLC) and Montreal Locomotive Works. The first main production batch of 300 locomotive was manufactured by them in 1949. They were numbered from 7216 to 7515. A total of 100 locomotive was built by Canadian Locomotive Company in 1955-56. Another 60 was produced by Fabryka Lokomotyw, of Chrzanów, Poland, and Lokomotivfabrik Floridsdorf of Vienna, Austria in 1957-59. The next batch of 259 was produced in India by CLW from 1963 and 1966. They were 5 tons heavier and were classified as the WP/Is.

The WP Class locomotives, were the crack locomotives of the Indian Railways in the 1960s and 1970s and hauled some of Indian Railway's most prestigious trains like the Taj Express, Grand Trunk Express, Howrah-Madras Mail, Frontier Mail and the Air Conditioned Expresses. The locomotives were used well into the 1980s and in a limited manner were used till the 1990s. Most of them have been retired today and only nine of them are preserved all over India.

Some of them have been overhauled and have been made operational for mail line duties. They are used to operate special trains. Azad, bearing serial number 7200 is one of the most famous of the steam locomotives of Indian Railways, as it not only hauls special trains but has appeared in a number of movies as well. It was manufactured in 1947 by Baldwin Locomotive Works, USA.

Factsheet:-
- Fuel Type:- Coal.
- Manufacturer:- Baldwin Locomotive Works, Canadian Locomotive Company, Montreal Locomotive Works, Fabryka Lokomotyw, Lokomotivfabrik Floridsdorf and Chittaranjan Locomotive Works.
- Production Period:- 1947 to 1967.
- Numbers Produced:- 755.
- Whyte:- 4-6-2.
- Tractive Effort:- 136.12 kN.
- Rated Power:- 2,680 HP.
- Top (Rated) Speed:-.
- Weight:- 113.7 Tons.
- Axle Load:- .
- Serial Numbers:- 7000 to 7754.

Class WG

The WG Class was a 2-8-2 configured locomotive and is the most produced steam locomotive of Indian Railways. About 2,450 of these locomotives were produced from 1950 to 1970. They were manufactured by a number of manufacturers including Chittaranjan Locomotive Works in India. The WG Class was introduced in 1950 and used almost the same equipment as the WP Class of Locomotive. The first hundred locomotives was built by North British and of the hundred, ten was subcontracted to Vulcan Foundry. CLW initially produced the locomotives with imported parts but by 1953, it was able to source 70% of the locomotive parts from the country. The locomotive was also supplied by Franco Belge of Raismes, France, Germany, Austria, Italy and Japan before they were fully produced by CLW.

The locomotive had a water capacity of 5,000 gallons and coal capacity of 18 tons. Many parts of the locomotive was interchangeable with the US design WP locomotive.

The production of the locomotive was stopped in 1970 and the last unit was named as the 'Antim Sitara' with the serial number of 10560. Today none of them remain in active service and most of them have been scrapped but eight of them have been preserved.

Factsheet:-
- Fuel Type:- Coal.

- Manufacturer:- Chittaranjan Locomotive Works, North British Locomotive Company, Vulcan Foundry, Anglo-Franco-Belge (La Croyère), Henschel, Gio. Ansaldo & C, Baldwin Locomotive Works, Henschel & Sohn, Hitachi, Krupp and Lokomotivfabrik Floridsdorf.
- Production Period:- 1950 to 1970.
- Numbers Produced:- 2,450.
- Whyte:- 2-8-2.
- Tractive Effort:- 172.99 kN.
- Rated Power:- 2,680 HP.
- Top (Rated) Speed:-.
- Weight:- 176.4 Tons.
- Axle Load:- .
- Serial Numbers:- .

WL Class (1939)

The WL Class of locomotive was a 4-6-2 configured light locomotive, with an axle load of about 16 tons. They were produced in 1939 and share the same nomenclature with another locomotive which was produced in 1955. While the two Class of locomotives share the same nomenclature, they do not have any similarity. A total of 4 of such locomotives were built by Vulcan Foundry for the North Western Railways. These were later transferred to Pakistan during partition.

Factsheet:-

- Fuel Type:- Coal.
- Manufacturer:- Vulcan Foundry.
- Production Period:- 1939.
- Numbers Produced:- 4.
- Whyte:- 4-6-2.
- Serial Numbers:- 101 to 104.

WL Class (1955)

The WL Class of locomotive of 1955 was also a 4-6-2 configured light locomotive, but with a higher axle load of 16.75 tons. They were produced in two batches, the first being of 10 locomotives in 1955 by Vulcan Foundry. The second batch of 94 was produced in India by Chittaranjan Locomotive Works between 1966 and 1969. The ten locomotive which were imported from UK were evenly distributed between the Northern and the Southern Railway.

The boiler of the locomotive consisted of a three-course barrel and had a round-topped firebox with the combustion chamber. The water tender tank had a capacity of 4,500 gallons and its coal capacity was 12 tons.

The locomotives were not transferred during partition and all served the Indian Railways till 1995. They operated on the branch lines around Firozpur, Jallandhar, Godhra, Rajahmundry and Shoranur. Today all of them have been retired and scrapped except the number 15005, which has been preserved.

Factsheet:-
- Fuel Type:- Coal.
- Manufacturer:- Vulcan Foundry and CLW.
- Production Period:- 1955 to 1969.
- Numbers Produced:- 104.
- Whyte:- 4-6-2.

- Tractive Effort:- 122.95 kN.
- Weight:- 98.7 Tons.
- Axle Load:- 16.75 Tons.
- Serial Numbers:- 15000 to 15009 and 15014 to 15107.

WM Class

The WM Class is a 2-6-4T tank locomotive and was introduced in India in 1942. A total of 70 of them were imported and later 4 WV Class locomotives were also converted to this class. Thirty of these locomotives were produced by Robert Stephenson and Hawthorns and the rest including the conversions was done by Vulcan Foundry. Ten of them were imported from Vulcan Foundry in 1942 and were used by GIPR and Eastern Indian Railway. The second batch of forty locomotives from Vulcan Foundry arrived in 1952 and was used by Eastern Indian Railways for heavy suburban duties near Calcutta.

The boiler of the locomotive had a Belpaire firebox and the inner shell was a all welded steel shell with two arch tubes. The locomotive had a water capacity of 3,000 gallons and coal capacity of 6.5 tons.

Factsheet:-
- Fuel Type:- Coal.
- Manufacturer:- Robert Stephenson & Hawthorns & Vulcan Foundry.
- Production Period:- 1939 to 1954.
- Numbers Produced:- 70.
- Tractive Effort:- 84.96 kN.
- Weight:- 108.2 Tons.
- Axle Load:- 16.25 Tons.
- Serial Numbers:- 13000 to 13073.

WT Class

The WT Class locomotive was a 2-8-4 T configured heavy tank locomotive with an axle load of 19 tons. It was designed and manufactured in India by Chittaranjan Locomotive Works in 1959. They were produced to haul heavy suburban trains and were based on the WM Class. The locomotives proved quite successful and hauled 12 coach suburban trains as compared to 8 coaches pulled by the WM Class. They continued to work, till the line was electrified. The initial orders were for 146 such locomotives but production was stopped at 30 when the lines were electrified. The first 10 went to Calcutta and the next 20 went to Madras. The last WTs operated near Rajahmundry in early 1980s.

Factsheet:-

- Fuel Type:- Coal.
- Manufacturer:- Chittaranjan Locomotive Works.
- Production Period:- 1959 to 1967.
- Numbers Produced:- 30.
- Whyte:- 2-8-4 T.
- Axle Load:- 18 Tons.
- Serial Numbers:- 14000 to 14029.

WU Class

The WU Class was a 2-4-2T configured tank locomotive and had an axle load of 16.5 Tons. It was produced by Vulcan Foundry for Indian Railways. A total of 4 was built by them from 1940 to 1943. It was proposed to procure additional numbers of this class, but its power was found to be packing to the requirement and as such only 4 of them ever ordered.

Factsheet:-

- Fuel Type:- Coal.
- Manufacturer:- Vulcan Foundry.
- Production Period:- .
- Numbers Produced:-
- Whyte:- .
- Tractive Effort:- .
- Rated Power:- .
- Top (Rated) Speed:-.
- Weight:- Tons.
- Axle Load:- 16.5 Tons.
- Serial Numbers:- 26950 to 26953.

WV Class

The WV Class of locomotives are 2-6-2T locomotives with a axle load of about 16.25 tons. Like the WU Class, only 4 of them were procured from Vulcan Foundry, UK and they too were later converted into the WM Class with 2-6-4T configuration.

Factsheet:-

- Fuel Type:- Coal.
- Manufacturer:- Vulcan Foundry.
- Production Period:- 1939 to 1942.
- Numbers Produced:- 4.
- Whyte:- 2-6-2T.
- Tractive Effort:- .
- Rated Power:- .
- Top (Rated) Speed:-.
- Weight:-   Tons.
- Axle Load:- .
- Serial Numbers:- 13070 to 13073.

WW Class

The WW Class was a 0-6-2T configured class of locomotive and had an axle load of 16.5 Tons. They were produced by Vulcan Foundry from 1940 to 1942. They were the smallest of the W family of locomotives and were used for shunting duties at large stations and marshalling yards. The locomotives were originally deployed with the North Western Railway but after partition they were transferred to Northern Railway. They were deployed in the Delhi station and yard.

Factsheet:-

- Fuel Type:- Coal.
- Manufacturer:- Vulcan Foundry.
- Production Period:- 1940 to 1942.
- Numbers Produced:- 4.
- Whyte:- 0-6-2T.
- Tractive Effort:- 87.67 Tons.
- Rated Power:- .
- Top (Rated) Speed:-.
- Weight:- 73.75 Tons.
- Axle Load:- .
- Serial Numbers:- 15010 to 15013.

Meter Gauge Locomotives

Most meter gauge steam locomotive used by the Indian Railways were imported but in 1953 and so, it was decided to source most of them from Tata Group. The Singbham workshop of the Easten Indian Railways was sold to the Tata Group on the 1st of June 1945 and TELCO was formed to take over the workshop and manufacture locomotives. The TELCO was initially given a target of 50 locomotives per year and this was increased to 75 and then to 100 per year and it was able to meet all the targets. By 1958-59, TELCO was producing more than 100 locomotives per year and in fact produced 103 such locomotives that year. In fact an agreement was signed by which TELCO was to produce such locomotives for Indian Railways for the Third Five Year Plan period. The workshop also specialized in the manufacture of SGS Class boilers.

The types of meter gauge steam locomotives was used by Indian Railways and they are as follows:-

Nilgiri Mountain Railway X class

BESA designs:
1. 4-6-0 Passenger
2. 4-6-0 Mixed
3. 4-8-0 Goods
4. 2-6-2T Tank

Wartime designs:
1. Class MAWD: 2-8-2 USATC S118 Class
2. Class MWGX: 4-6-2+-6-4 Garratt

Indian Railway Standards designs
1. Class YA: 4-6-2 with 9-ton axle load
2. Class YB: 4-6-2 with 10-ton axle load
3. Class YC: 4-6-2 with 12-ton axle load
4. Class YD: 2-8-2 with 10-ton axle load
5. Class YE: 2-8-2 with 12-ton axle load (none built)
6. Class YF: 0-6-2; later examples were 2-6-2
7. Class YK: 2-6-0 version of the 2-6-2 YF
8. Class YL: 2-6-2
9. Class YT: light 0-4-2T
10. Class YG: 2-8-2 goods locomotive
11. Class YP: 4-6-2 passenger locomotive

Nilgiri Mountain Railway X class

The Nilgiri Mountain Railway X Class was a 0-8-2T rack and pinion compound locomotive. They were used on the 20 Kms section between Coonoor and Mettupalayam of the Nilgiri Mountain Railways. The locomotive was first acquired to replace the Beyer-Peacock 2-4-0 engines which were used on the line but were found to be underpowered for the purpose. They were brought in two batches Swiss Locomotive and Machine Works, Winterthur, Switzerland, with the first batch of 12 being imported from 1920 to 1925 while the second batch of 5 was delivered in 1952. A further 4 of this Class manufactured by Golden Rock Railway Workshop from 2011 to 2014. The 4 new locomotives are oil fired locomotives.

The locomotives are equipped with two high-pressure and two low-pressure cylinders and they are located outside the locomotive's frames. In 2002, the old coal fired locomotives numbering 37395 was modified to use oil and another was also slightly modified to do so. The shift from coal to fuel oil was done to minimize the risk of forest fire and the ease of refueling with fuel oil.

Factsheet:-

- Fuel Type:- Coal and fuel oil.
- Manufacturer:- Swiss Locomotive and Machine Works, Winterthur, Switzerland and Golden Rock Railway Workshop, Tiruchirappalli,.
- Production Period:- 1914 to 2014.
- Numbers Produced:- 21.
- Whyte:- 0-8-2T.

Other Locomotives

The YB Class of locomotive was a 4-6-2 configured meter gauge locomotive with an axle load of 10-ton. The Ajmer workshop of BB&CI produced a total of 58 of these locomotives. The boiler was supplier by Tata from its TELCO works at Singbhum.

The YP Class of locomotive was a 4-6-2 configured passenger locomotive. The design locomotive was later improved and an order for 150 of them was made in 1951-52, with delivery scheduled by middle of 1952. The YG Class was a 2-8-2 configured goods locomotive. One such locomotive bearing the serial number YG 4119 has been preserved and exhibited before the Guwahati Railway Station. The locomotive was manufactured in 1956 in Germany and was stationed at Badarpur station of Lumding Division. It retired in 1997 after 41 years of service.

Narrow Gauge Locomotives

Indian Railways has two types of narrow gauge railway, one is 2 feet 6 in and the other is 2 feet in width. The locomotives used by the Indian Railways on the narrow gauge routes are as follows:-

2 ft 6 in

Barsi Light Railway:

1. Class A: 0-8-4T
2. Class B: 4-8-4T
3. Class C: 0-6-0ST
4. Class D: 0-4-0
5. Class E: Sentinel railcars
6. Class F: 2-8-2
7. Class G: 4-6-4

Indian Railway Standards:

1. Class ZA: 2-6-2 with 4.5-ton axle load (none built)
2. Class ZB: 2-6-2 with 6-ton axle load
3. Class ZC: 2-8-2 with 6-ton axle load (none built)
4. Class ZD: 4-6-2 with 8-ton axle load (none built)
5. Class ZE: 2-8-2 with 8-ton axle load
6. Class ZF: 2-6-2T with 8-ton axle load

2 ft

Darjeeling Himalayan Railway:

1. DHR A Class;
2. DHR B Class.
3. DHR C Class.
4. DHR D Class.

Indian Railway Standards (none built):

1. QA: 2-6-2 with 4.5-ton axle load
2. QB: 2-6-2 with 6-ton axle load
3. QC: 2-8-2 with 6-ton axle load

DHR A Class Locomotives

The DHR A Class Locomotive was a 0-4-0 WT well tank steam locomotive and was used on the 2 ft narrow gauge line. It was the first standard design locomotive of the Darjeeling Railway. The locomotive was manufactured by Sharp, Stewart & Co. and Hunslet in 1881. A total of 8 locomotive of this Class was used and the first two and last two were produced by by Sharp, Stewart & Co., while the rest were produced by Hunslet. It was a 12 ton locomotive and had a water capacity of 250 gallon capacity. The locomotives were prone to derailment but the problem overcome by distributing the weight more evenly and ensuring a lower centre of gravity.

Today all of them have been withdrawn in 1954 and most have been scrapped. Only one locomotive of this Class has been preserved at Tindharia.

Factsheet:-
- Fuel Type:- Coal and fuel oil.
- Manufacturer:- Sharp, Stewart & Co. and Hunslet, UK.
- Production Period:- 1881.
- Numbers Produced:- 8.
- Whyte:- 0-4-0 WT.
- Top (Rated) Speed:- 30 kmph.
- Weight:- 12 Tons.
- Serial Numbers:- 9 to 16.

DHR B Class Locomotives

The DHR B Class Locomotive is a famous 0-4-0 ST saddle tank steam locomotives. A total of 34 was manufactured from 1889 to 1925 by Sharp, Stewart and Company, North Britsh Locomotive Co., Baldwin Locomotive Works and DHR Tindharia Works. The locomotives had a weight of 14 tons and a water capacity of 400 gallons.

The locomotives served the railways well and for more than a hundred years. Today only a few of these locomotives are in service with the Darjeeling Railway as four was sold to Coal India, Assam and one was sold for private preservation. While some of them have been retired and scrapped, two of them i.e. the 'Tusker' and the 'Victor' haul trains regularly between Darjeeling to Ghum stations via Batasia loop.

Factsheet:-

- Fuel Type:- Coal and fuel oil.
- Manufacturer:- Sharp, Stewart & Co. and Hunslet, UK.
- Production Period:- 1881.
- Numbers Produced:- 8.
- Whyte:- 0-4-0.
- Top (Rated) Speed:- 30 kmph.
- Weight:- 14 tons.

DHR C Class Locomotive

The DHR C Class was a 4-6-2 configured locomotive. Two locomotives of this Class was built in 1914 and was mainly used on the plain section of the line. While it was intended to produce about twice the power as compared to a DHR B Class locomotive, it was only able to produce only 65% more power.

The locomotives were retired in November 1954 and both have been preserved.

Factsheet:-
- Fuel Type:- Coal.
- Manufacturer:- Beyer, Peacock and Company.
- Production Period:- 1911.
- Numbers Produced:- 2.
- Whyte:- 4-6-2.

DHR D Class Locomotive

The DHR D Class was a 0-4-0+0-4-0 Garratt-type articulated steam locomotive. It was essentially two B Class locomotive put together. Only one locomotive of this class was ever built but the project was shelved due to the limited success it achieved.

Production Units

Production Units

The Steam locomotives were the first locomotives to be used in the Indian Railways and most of them were imported from manufacturer in UK. As the demand for locomotives increased, the number of supplier also increased and in course of time locomotives were imported from a number of countries. It was only with the setting up of the Chittaranjan Locomotive Works in 1950, that manufacturing of steam locomotives started in India. The major units producing and modifying steam locomotives for Indian Railways are as follows:-

1. Chittaranjan Locomotive Works, Chittaranjan.
2. Vulcan Foundry, UK.
3. North British Locomotive Company, UK.
4. Beyer, Peacock and Company, UK.
5. Robert Stephenson and Hawthorns, UK.
6. William Beardmore & Co, UK.
7. Hunslet, UK.
8. Swiss Locomotive and Machine Works, Switzerland.
9. Canadian Locomotive Company.
10. Montreal Locomotive Works.
11. Fabryka Lokomotyw.
12. Lokomotivfabrik Floridsdorf.
13. Baldwin Locomotive Works
14. Railway Workshops & Others.

Chittaranjan Locomotive Works

The Chittaranjan Locomotive Works located in Chittaranjan near Asansol in West Bengal. It is the oldest locomotive production unit in India. It was set up with an initial capacity of producing 50 steam locomotives per year along with 50 additional boilers, in collaboration with North British Locomotive Company. While today it is a dedicated electric locomotive manufacturing unit but it started production with steam locomotives. Some of the steam locomotives produced by it were the WP and the WG Class Steam Locomotives. The construction began in April 1948 and it was named as Chittaranjan Locomotive Works, in honor of the freedom fighter Deshbandhu Chittaranjan Das in the budget of 1949-50.

It was founded on the 26th of January 1950 and the first locomotive produced by it was a WG Class locomotive on the 1st of November 1950. It was also named after the freedom fighter Deshbandhu Chittaranjan Das, with the serial number of 8401. The locomotive was flagged off by the then President of India Dr Rajendra Prasad.

The first 10 locomotives at CLW were not manufactured but only assembled by them from imported parts. The specialized equipment for manufacturing steam locomotives was initially obtained from Vulcan Foundry, UK. The production slowly picked up pace and with it the percentage of indigenous parts also went up. By 1953, 70% of all parts was sourced domestically. CLW produced its 100th locomotive on the 6th of January 1954. At the peak of production of steam locomotives in India, it produced about 168 broad gauge steam engines per year. This was from 1956 to about 1966. The fist WP Class locomotive to be produced was in February 1963 and were named as the WP1 Class.

It manufactured a total of 2,351 steam locomotives from 1950 to 1972, which included 94 WL Class, 30 WT Class, 259 WP Class and about 1,900 WG Class Locomotives. While it continued producing steam locomotives, its first diversification was seen in 1961 when it converted a steam workshop to an electric one and took up production of 12 DC locomotives. The production of steam locomotives was stopped in 1971, when it was decided to stop domestic production of broad gauge steam locomotives. The last WP locomotive to be built was in 1967, bearing the number 7754 and the last steam locomotive to be produced here was a WG Class locomotive rechristened as the "Antim Sitara.

Vulcan Foundry

The Vulcan Foundry Limited, was a British locomotive company and it supplied some of the earlier steam locomotives to Indian Railways. The company was located at Newton-le-Willows in Lancashire and operated between 1832 and 1962. The company was founded in 1832 as the Charles Tayleur and Company to produce girders for bridges, switches, crossings, and other ironwork. In 1847, it was renamed as the Vulcan Foundry Company and in 1898, it became the Vulcan Foundry Limited. The company was a pioneer is steam locomotives and later went on to produce a number of diesel and electric locomotives also.

The company supplied a number of steam locomotives to the Indian Railways and was one of the main suppliers of steam locomotives to India. The locomotives supplied were the XA, XB, XC, XE, XP, XS, WM, WV, WL, WG, WW, WU and the WG Class of locomotives. It supplied a total of about 260 steam locomotives to the Indian Railways and was also the most diverse supplier, given the number of different locomotives supplied. The XA, WM and the XE Class were supplied in large numbers, as about 113 of the XAs, 49 of the XEs and 40 of the WMs were supplied over time. It also supplied about 14 of the WL Class, 10 of the WG Class and 4 each of the WU, WW, XS and WV Class of locomotives.

The company was passed onto to number of giant corporations namely GEC, Alstom and MAN Diesel and in 2002, the works was finally closed. This brought an end to one of the glorious pioneers of locomotives in the World. In 2007, all structures of the works were demolished to pave way for building new houses and today not even a single structure exists of the old locomotive works but the number of locomotives which it has built and which are still preserved all over the World would always remain as testimony of the achievements of the company and as such the nostalgia behind the company is great and number of websites exist which display information relating to its past including many editions of it magazine tilted "Vulcan".

North British Locomotive Company

The North British Locomotive Company, was one of the largest locomotive manufacturer of its time and supplied the Indian Railways with a number of classes of locomotives. The company was created in 1903, by the merger of three locomotive manufacturing companies namely Sharp, Stewart and Company, Neilson, Reid and Company and Dübs and Company. The company started its production with steam locomotives and later diversified to diesel and electric locomotives also.

The company supplied a number of locomotives to India and they were the E1, XB, SG and WG Class of locomotives. While the E1 was a custom built locomotive and was specially built for GIPR, the rest were standard design ones. About 115 of the WG Class and 4 of the XB Class was supplied to India by them. The DHR A and DHR B Class of meter gauge steam locomotives used by the Darjeeling Himalayan Railway was supplied by Sharp, Stewart and Company. It was one of the predecessor of the company.

The advent of the diesel and electric locomotives sounded the death kneel for the company, as its attempt to produce the new locomotives failed and had to be withdrawn from service very shortly. The company went under liquidation in April 1962 and its assets sold to different entities. The goodwill of the company was brought by Andrew Barclay Sons & Co. Today the number of preserved locomotives built by it, stand testimony of the works and ingenuity of the company.

Beyer, Peacock and Company

The Beyer, Peacock and Company, was also an English locomotive manufacturing company and supplied a number of locomotives to Indian Railways. The company was located in Gorton, Manchester and was founded in 1854 by Charles Beyer, Richard Peacock and Henry Robertson. The company mainly produced steam locomotives but also later attempted to produce diesel and electric locomotives.

The Company produced the Beyer Garratt types of locomotives, which are locomotives with three parts. The boiler is mounted on the centre frame, with the two steam engines mounted on separate frames on each end of the boiler.

The company supplied a number of locomotives to the Bengal Nagpur Railway, the Nilgiri Mountain Railway and the Darjeeling Himalayan Railways. The Bengal Nagpur Railway was supplied with the M, HSG, N, NM and P class of locomotives. While the Nigiri Mountain Railway got their X Class and Darjeeling Himalayan Railway got the DHR D Class of locomotives from them.

The company remained in operation till the 1960s and was closed down in 1962. Today many of its locomotives remained preserved all over the World.

Robert Stephenson and Hawthorns

The Robert Stephenson and Hawthorns was a locomotive builder located in North Eastern England. The Company was formed in 1937, when Robert Stephenson and Company took over the locomotive building department of Hawthorn Leslie and Company. It built a number of steam, diesel and electric locomotives over the years and it supplied the SG and WM Class of steam locomotives of the Indian Railways.

The company remained operational till it was closed in 1964 but today its memories remains in the numerous steam locomotives which are persevered at various locations of the World.

William Beardmore & Co

The William Beardmore & Co, was a Scottish Engineering and shipbuilding conglomerate. It was based in Glasgow and was founded in 1893 and operated till 1983. It was founded by William Beardmore, The Company made vehicles for all mode of transportation such as planes, ships, cars and locomotives. It supplied a number of locomotives to India which included both custom made locomotive and standard design locomotives.

It supplied 20 HGS Class, 3 LS Class, 2 XC Class and 44XE locomotives to the Eastern Indian Railways. 25 Class H, 4 XC Class and 2 BE Class locomotives was supplied to the Bombay, Baroda and Central India Railway. 2 W Class Locomotives were supplied to the Madras and Southern Mahratta Railway. 5 HPS Class, 26 XC Class and 2 DL Class locomotives were supplied to the North Western Railways.

The Company closed its operations 1983 with the closure of the Parkhead Forge.

Hunslet Engine Company

Hunslet Engine Company, is an Engineering company located in Leeds, England and specializes in the manufacturing of shunting locomotives. The company mainly produced steam shunting locomotives but later shifted to diesel shunting locomotives. The Company now has a subsidiary known as 'The Hunslet Steam Company' and is specializes in the maintenance and manufacture of steam locomotives.

The company was founded by John Towlerton Leather in 1864 in Hunset, England. The company supplied the DHR A and DHR B Class of narrow gauge locomotive to the Darjeeling Railways. Today is among a few steam locomotive producing companies which still in operation.

Swiss Locomotive and Machine Works

The Swiss Locomotive and Machine Works, was a railway equipment manufacturing company and was based in Winterthur, Switzerland. It was a specialist in the manufacturing of equipment for mountain railways and manufactured for almost all mountain railways of the World. The Company was founded in 1871 by British Engineer Charles Brown. It produced both steam and electric locomotives.

The Company supplied the first batch of locomotives to the Nilgiri Mountain Railways, which are known as the Nilgiri Mountain X Class Locomotives. They also supplied the Indian Railways with its first electric locomotives which were DC powered and were known as the WCG class locomotives. The details of this locomotive can be found in my other book titled 'DC Locomotive of the Indian Railways'. The company was closed in 2005, when its last part was also sold off.

Canadian Locomotive Company

The Canadian Locomotive Company, was a manufacturer of locomotives and was located in Ontario, Canada. The company was founded in February 1978 and continued its operation till April 1969. The Company manufactured both steam and diesel locomotives and produced more than 3,000 locomotives during its operational years. It was Canada's second largest locomotive builder after Montreal Locomotive Works. It supplied the Indian Railways with the WP Class of Steam locomotives. A total of about 200 such locomotive was imported from them to India.

The company was closed in April 1969 due to declining income and strikes. Their works was demolished in August 1971.

Montreal Locomotive Works

The Montreal Locomotive Works, was Canada's biggest manufacturer of locomotive and was located in Montreal, Quebec. The company was founded in 1888 and continued its operations till 1985.

The Company was founded in 1888 to cater to the demand of locomotives in Canada. In 1901, the company merged itself with other locomotive manufacturing companies to form the American Locomotive Company (ALCO). The company remained a subsidiary of ALCO for many years.

The first diesel locomotives of the Indian Railways was in fact supplied by ALCO and till recently most of the diesel locomotives of Indian Railways had ALCO looks only. The details of this can be read in my other book titled "Diesel Locomotives of the Indian Railways".

The company supplied India with the WP Class of Steam locomotives and a total of 120 them was imported from them. The Company was closed in 1985 and the plant was sold to Bombardier Transportation in 1988.

Fabryka Lokomotyw

The Fabryka Lokomotyw, is a Polish locomotive manufacturer and is located in Chrzanów. The company was founded in 1919 as the Pierwsza Fabryka Lokomotyw w Polsce Sp. Akc. meaning 'First Factory of Locomotives in Poland Ltd'. The company was nationalized in 1947 and named as Fablok. In 2001, it was privatized to form a Joint stock company. In 2013, the company declared bankruptcy and was brought by Martech Plus. It does not produce new locomotives.

The company supplied Indian Railways with 30 numbers of WP Class Steam locomotives.

Lokomotivfabrik Floridsdorf

The Lokomotivfabrik Floridsdorf, was an Austrian Locomotive Company and was the third locomotive producing company in Austria. It was founded in September 1869.

It supplied the Indian Railways with 30 nos. of the WP Class and 60 nos. of WG Class locomotives.

Baldwin Locomotive Works

The Baldwin Locomotive Works, was an American locomotive manufacturing company and was located in Eddystone, Pennsylvania. The company was founded in 1825 by Matthias W. Baldwin and was in production till 1956. It produced over 70,000 locomotives during it years in operations and was finally closed in 1972. It was the largest manufacturer of steam locomotives in the World for many years but the switch to diesel locomotives ended its glorious run.

The company supplied the Indian Railways with 116 nos. of WP Class and 50 Nos. of WG Class of locomotives.

Railway Workshop

A number of Railway workshops, also gained in manufacturing capability while performing the tasks of repair and maintenance. In fact prior to Independence, the only manufacturing of locomotive in India was done the Ajmer Workshop of BB&CI. The workshop had built its first locomotive in 1896, which was a 0-6-0 tender locomotive and after that it mostly produced meter gauge locomotives. In fact, it produced about more than 400 meter gauge locomotives. It also produced about 20 XT Class locomotives in 1947 but sadly manufacturing activity in the workshop was ceased after the production of the XT Class locomotives and it focused only on repairs after that. .

The Golden Rock workshop, is today one of the finest diesel workshops of the Indian Railways. It is located in Tiruchirappalli, Tamil Nadu and I have described it in details in my book on diesel locomotives. The workshop also has a strong manufacturing setup and produces the steam locomotive for the Nilgiri Mountain Railways, which are named as the NMR X Class.

Other Manufacturers

Apart from the manufacturers mentioned, a number of other manufacturers have also supplied steam locomotives to the Indian Railways and they are as follows:-

1. Anglo-Franco-Belge (La Croyère)
2. Henschel, Gio. Ansaldo & C
3. Henschel & Sohn.
4. Hitachi.
5. Krupp.
6. Winterthur, Switzerland.

While the first five manufacturers supplied India with the WG Class of locomotive, the last one provided India with the Nilgiri Mountain X Class.

Steam Locomotive Sheds

Indian Railways had a number of steam locomotive sheds to take care of the maintenance and overhaul of its fleet of steam locomotives. During the best days of steam traction in India, hundreds of such sheds were present all over India but sadly today only one such shed namely the Reawri Locomotive shed is still in existence. The shed has been the location for many movie shootings such as Veer Zaara, Gadar etc and hence is quite popular.

The working of steam locomotives were quite apart from the other types of locomotive as it required a lot of maintenance and repairs. This was due to the large number of moving parts in a steam locomotive and also, the fact that they had a very limited range. The range also limited due to the fact that the locomotive had to carry with it coal and water. While water could be filled at intermediate station with the help of water columns along the tracks, the loading of coal required that the locomotive be disconnected and sent to a centralized depot with specialized facility. It was due to these reasons that the number of sheds for steam locomotives were far more than for diesel or electric locomotives.

These sheds were different from the central workshops which had full facilities to repair and overhaul locomotives. The sheds were equipped to perform the work of routine maintenance and repairs and also had adequate spares for such work. The routine maintenance was done as per fixed schedules for each type and class of locomotive, which was based on the distance travelled by the locomotive. The sheds were of two types i.e. home sheds and turn round sheds as per the scope of work to be done. The home sheds were the ones to which the locomotives were assigned to and had full facilities for service and repairs. The turn round sheds had limited facilities of repairs and as such were equipped to perform minor repairs only. They were not allocated any locomotives.

The first repair facilities to come up in India were in form of workshops and the first one was set up by GIPR at Byculla in 1854 and this was followed by one in Howrah, set up by East India Railway. BB&CCI set up its first workshop at Amroli in 1856. With the spread of the railway network, it became apparent that the centralized locomotive workshops were quite inadequate to cater to the needs of such a vast network and also, it was not economical to build large number of workshops. So, the locomotive sheds were set up and soon they were built across the length and breadth of the country.

The different sheds under different Railway zones which were once operational, are as follows:-

1. Central Railway (Broad Gauge)
    a. Agra Cantt.
    b. Ajni
    c. Amla
    d. Badnera
    e. Bhusaval
    f. Bina

g. Dehu Road
   h. Igatpuri
   i. Itarsi
   j. Jabalpur
   k. Jhansi
   l. Kalyan
   m. Lonavla
   n. Matunga Workshops
   o. Nandgaon and Chalisgaon
   p. New Katni Jn.
   q. Parel Workshops
   r. Pune
   s. Satna
   t. Wardha
2. Central Railway (Meter & Narrow Gauge)
   a. Akola
   b. Khandwa
   c. Daund
   d. Dhaulpur
   e. Gwalior
   f. Kurduwadi
   g. Murtajapur
   h. Neral
   i. Pachora

    j. Pulgaon
3. Western Railway (Broad Gauge)
    a. Anand
    b. Bandra
    c. Bayana
    d. Gangapur
    e. Godhra
    f. Idgah (for Agra)
    g. Kankaria
    h. Kota
    i. Parel (for Bombay)
    j. Ratlam
    k. Sawai Madhopur
    l. Udhna
    m. Ujjain
    n. Vadodara
    o. Valsad
    p. Viramgam
4. Western Railway (Meter Gauge)
    a. Abu Road
    b. Ajmer
    c. Bandikui
    d. Bhavnagar and Dhola sub-shed
    e. Gandhidham

f. Hapa
g. Jaipur
h. Jetalsar
i. Junagadh
j. Khambli Ghat
k. Mahesana
l. Marwar Jn.
m. Mavli Jn.
n. Mhow
o. Nimuch
p. Palanpur
q. Phulera
r. Rajkot
s. Sabarmati
t. Surendranagar
u. Udaipur (Rana Pratap Nagar, RPZ)
v. Viramgam
w. Wankaner

5. Western Railway (Narrow Gauge)
    a. Ankleshwar
    b. Bhavnagar
    c. Bilimora
    d. Dabhoi
    e. Devgadh Bariya

- f. Godhra
- g. Halol
- h. Kosamba
- i. Morvi
- j. Nadiad
- k. Pratapnagar.
- l. Bondamunda.
- m. Bokaro.
- n. Kharagpur.

6. Eastern Railway (Broad Gauge)
   - a. Andal
   - b. Asansol
   - c. Bandel
   - d. Barkakana
   - e. Barwadih
   - f. Burdwan
   - g. Chitpur
   - h. Chopan
   - i. Danapur
   - j. Dhanbad
   - k. Garhara
   - l. Gaya
   - m. Gomoh
   - n. Howrah

    o. Jamalpur
    p. Jhajha
    q. Katrasgarh
    r. Madhupur
    s. Mughalsarai
    t. Naihati
    u. Pathardih
    v. Patratu
    w. Rampurhat
    x. Ranaghat
    y. Sahibganj
    z. Sitarampur
    aa. Sonnagar

7. Northern Railway (Broad Gauge)
    a. Allahabad
    b. Amritsar
    c. Bareilly
    d. Bhatinda
    e. Delhi Jn.
    f. Faisabad
    g. Firozpur
    h. Ghaziabad
    i. Jalandhar Cantt.
    j. Jind

- k. Kalka
- l. Kanpur
- m. Laksar Jn.
- n. Lucknow
- o. Ludhiana
- p. Moradabad
- q. Pathankot
- r. Pratapgarh
- s. Roza
- t. Saharanpur
- u. Tughlakabad
- v. Tundla
- w. Varanasi

8. Northern Railway (Meter Gauge)
   - a. Bikaner
   - b. Churu
   - c. Delhi Jn.
   - d. Hanumangarh
   - e. Jodhpur
   - f. Merta Road
   - g. Rewari
   - h. Samdari
   - i. Sirsa
   - j. Unallocated

9. Northern Railway (Narrow Gauge)
    a. Pathankot
    b. Kalka.
10. South Central Railway (Broad Gauge)
    a. Bitragunta
    b. Daund
    c. Dornakal
    d. Ghorpuri
    e. Kazipet
    f. Lallaguda
    g. Miraj
    h. Rajahmundhry
    i. Vijayawada
    j. Wadi
11. South Central Railway (Meter Gauge)
    a. Castle Rock
    b. Donakonda
    c. Gadag
    d. Hubli
    e. Lallaguda
    f. Miraj
    g. Nandial
    h. Purna
    i. Tadepalli

12. South Eastern Railway (Broad Gauge)
    a. Adra
    b. Bhilai
    c. Bhojudih
    d. Bilaspur
    e. Bondamunda
    f. Chakradharpur
    g. Dangoaposi
    h. Dongargarh
    i. Hatia
    j. Jharsuguda
    k. Kharagpur
    l. Khurda Road
    m. Mahendragarh
    n. Nagpur
    o. Sahdol
    p. Santragachhi
    q. Waltair
13. South Eastern Railway (Narrow Gauge)
    a. Bankura
    b. Baripada
    c. Chindwara
    d. Gondia
    e. Motibagh

- f. Nainpur
- g. Naupada
- h. Raipur
- i. Ranchi

14. Southern Railway (Broad Gauge)
    - a. Arakkonam
    - b. Bangalore Cantt.
    - c. Basin Bridge
    - d. Erode
    - e. Jolarpettai
    - f. Nandalur
    - g. Podanur
    - h. Quilon
    - i. Raichur
    - j. Renigunta
    - k. Shoranur
    - l. Tiruchirapalli
    - m. Tondiarpet (for Madras)

15. Southern Railway (Meter Gauge)
    - a. Arsikere
    - b. Bangalore
    - c. Chingleput
    - d. Coonoor
    - e. Golden Rock Workshops

f. Guntakal
   g. Madras Egmore
   h. Madurai
   i. Mayuram
   j. Mysore
   k. Pakala
   l. Palni
   m. Shimoga Town
   n. Tiruchirapalli
   o. Tirunelveli
   p. Villupuram
   q. Yeshwantpur
   r. Quilon
16. Southern Railway (Narrow Gauge)
   a. Yelahanka.
17. Northeastern Railway (Broad Gauge)
   a. Samistipur.
   b. Sonpur.
18. Northeastern Railway (Meter Gauge)
   a. Barari
   b. Bareilly City
   c. Bhatni
   d. Charbagh
   e. Chupra Kachery

f. Darbhanga
g. Fatehgarh
h. Garahara
i. Gonda
j. Gorakhpur
k. Izzatnagar
l. Kanpur (Anwarganj)
m. Kasganj
n. Kashipur
o. Kathgodam
p. Mailani
q. Mau Jn.
r. Narkatiaganj
s. Pilibhit
t. Saharsa
u. Samastipur
v. Sonepur
w. Thanabihpur
x. Varanasi

19. Northeastern Frontier Railway (Broad Gauge)
    a. Malda Town.
    b. New Bongaigaon.
    c. New Jaipaiguri.
20. Northeastern Frontier Railway (Meter Gauge)

    a. Tinsukia

    b. Mariani Junction

    c. Lumding Jn

    d. Rangapara North

    e. Alipurduar Jn.

21. Northeastern Frontier Railway (Narrow Gauge)

    a. Tindharia (Darjeeling Himalayan Railway)

Turntable

    The turntable, was an interesting piece of equipment which was found in all such sheds, workshops and also many railway stations. While it was quite popular in use during the steam era and was required for the first types of diesel and electric locomotives but with advancement of technology the turntable has been rendered obsolete and can no longer be seen in stations but only in museums.

    The turntable was to turn around railway rolling stock specially locomotives. The Wye was also used for such purpose but it required much more space and so was more costly. It was required for locomotive which could be operated in one direction only. It consisted of track supported on two parallel girders, suspended on a central pivot. Wheels were attached on the end of the girders and they moved over rails along the circumference of the pit. It was installed in a circular masonry pit, with a sloping bottom. This was done to allow rain water to drain out. It had locking bolts to fix the locomotive in place and had two or more tracks radiating out from it. The diameter of the turntable was built to accommodate the largest of locomotive used in that section. The diameter of turntables used by Indian Railways are as follows:-

1. Broad Gauge:- 30.5 mts, 22.9 mts and 19.8 mts.
2. Meter Gauge:- 19.8 mts.
3. Narrow Gauge:- 15.75 mts.

The turntables were very popular and could be seen in all terminal stations, workshops, yards, sheds and on many stations also. The advent of new age locomotives which can move in both direction rendered the turntable obsolete and relegated them to museums. As such all turntables have been mostly been dismantled but in 2012, it did see a revivable when it was announced that turntables would be installed in line 1 of the Mumbai Metro.

Conclusion

The Steam locomotives were the pioneer locomotives of the Indian Railways and they are the one who took the Railways to the length and breadth of India. In this book, we have looked at how these locomotives evolved with time and their history. The steam locomotives were the first locomotives to be seen by most Indians and hence, they have been able to etch a special place in the mind of the nation.

It was the steam locomotive which took the railways to every corner of India and remained the main mover of the nation for many decades but by the late 1960s, it was evident that they were growing obsolete and were replaced by diesel and AC locomotives. While the production was stopped in 1971, they contained to be operated on main lines till 1980s, when they were relegated to branch lines. The curtains on steam locomotive came down finally in 2000, when the last steam operated passenger service at Wankaner in Gujarat was stopped. This was a meter gauge branch line.

The retirement of steam locomotive did not end its legacy and its impact can be gauged by the fact that even after decade's people still associate the locomotive with them only. The nostalgia and romance of the steam locomotives has never died down and surprisingly continues to grow. This fact can be known from the fact that special heritage trains hauled by steam locomotives are increasing in numbers. Today many premier tourist trains like Palace on Wheels, Maharaja Express and others are hauled by the steam Locomotives.

Today the days of Steam Locomotives are long past gone and most of them have been scrapped but yet their memories and nostalgic feelings continue to grow as time passes. As technology continues to grow and newer technology would surely make even the present day technology obsolete, it would be foolish to wish that steam locomotives would ever be seen on active operations again. Yet they would never die out altogether and would remain as a remainder of the pioneering days Railways and also the days when train travel was truly romantic for both the passenger and the crew.

Fig:- P Class Locomotive of BB&CI at National Rail Museum, Delhi

Fig:- ST (707) of North Western Railway

Fig:- MTR No. 2

Fig:- XC Class Steam Locomotive

Fig:- Crane Locomotive.

Fig:- Locomotive of the Patiala State Monorail Trainways

Fig:- XT Class locomotive of Eastern India Railways

Fig:- P Class Steam Locomotive of BB&CI.

Fig:- M2 Class Steam Locomotive of BB&CI.

Fig:- M2 Class Steam Locomotive of BB&CI.

Fig: Turntable at National Rail Museum, New Delhi.

Fig:- YG 4119 MG steam locomotive at Guwahati Railway Station

Resources

I would like to thank the esteemed reader for reading my book and hoped you have liked it. I would like to hear from you all and readers can find me on facebook and linkedin. Below are some of the links to my other resoucres which the readers may find to be interesting. .

My YouTube Channel:- https://www.youtube.com/user/twahiralam

Wordpress Site:- twahiralam.wordpress.com

www.ingramcontent.com/pod-product-compliance
Lightning Source LLC
Chambersburg PA
CBHW030703220526
45463CB00005B/1888